A LOOK AT WEATHER

SNOW AND RAIN

BY KATE MIKOLEY

Gareth Stevens Publishing

CRASHCOURSE

Please visit our website, www.garethstevens.com. For a free color catalog of all our high-quality books, call toll free 1-800-542-2595 or fax 1-877-542-2596.

Library of Congress Cataloging-in-Publication Data
Names: Mikoley, Kate, author.
Title: Snow and rain / Kate Mikoley.
Description: Buffalo : Gareth Stevens Publishing, [2024] | Series: A look at weather | Includes bibliographical references and index.
Identifiers: LCCN 2022061114 | ISBN 9781538287354 (library binding) | ISBN 9781538287347 (paperback) | ISBN 9781538287361 (ebook)
Subjects: LCSH: Precipitation (Meteorology)--Juvenile literature. | Snow--Juvenile literature. | Rain and rainfall--Juvenile literature.
Classification: LCC QC920 .M55 2024 | DDC 551.57/7--dc23/eng/20221223
LC record available at https://lccn.loc.gov/2022061114

First Edition

Published in 2024 by
Gareth Stevens Publishing
2544 Clinton Street
Buffalo, NY 14224

Designer: Karen Quigley
Editor: Kate Mikoley

Photo credits: Cover, p. 1 Melinda Nagy/Shutterstock.com; cloud background pp. 2-3, 32-33 kzww/Shutterstock.com; series art (snow background)Max/Shutterstock.com; series art (warm/cold front symbols) NickJulia/Shutterstock.com; series art (cloud background) sirachainan/Shutterstock.com; series art (cartoon cloud) MimaCZ/Shutterstock.com; p. 5 Elizabeth Gonano/iStock.com; p. 7 ArtMassa/iStock.com; p. 9 peresanz/Shutterstock.com; p. 11 Jia Ning Chen/Shutterstock.com; p. 13 Anest/Shutterstock.com; p. 15 Alexey Kljatov/Shutterstock.com; p. 17 lisdiyanto suhardjo/Shutterstock.com; p. 19 Kawaikini/commons.wikimedia.org; p. 19 Kokee_Park_-_Pihea_Lookout,_Pihea_Trail,_Hanapepe_(502977)_(16946945606)/commons.wikimedia.org; p. 21 Alan_Tow/iStock.com; p. 23 José Antonio Luque Olmedo/iStock.com; p. 25 XiFotos/iStock.com; p. 27 Ababsolutum/iStock.com; p. 29 carballo/iStock.com; p. 30 brgfx/Shutterstock.com.

Printed in the United States of America

CPSIA compliance information: Batch #CS24GS: For further information contact Gareth Stevens at 1-800-542-2595.

CONTENTS

Words in the glossary appear in **bold** type the first time they are used in the text.

PRECIPITATION STATION

When water falls from the sky, it's called precipitation. If the water is in **liquid** form, it's rain. If it's **solid**, it may fall as sleet, snow, or **hail**. In this book, you'll learn about the rain and snow forms of precipitation.

Make the Grade

Sleet is rain that freezes or partly freezes as it falls to the ground.

AN IMPORTANT CYCLE

Precipitation forms in the clouds. This is part of the water cycle, or how water moves through Earth and its **atmosphere**. Most of Earth's water is in oceans. Some is in lakes and rivers. When this water gets heated by the sun, it **evaporates**.

Make the Grade

More than 96 percent of all the water on Earth is in the oceans.

Evaporated water, called vapor,
moves up into the atmosphere.
Lots of vapor gathers in clouds,
forming larger drops of water.
This is called condensation.
When the drops get heavy enough,
they fall to the ground as rain.

Make the Grade

When rain falls to Earth, it may go back into oceans, rivers, or lakes. Or, it may collect in soil. No matter where it goes, it stays part of the water cycle.

PRECIPITATION PIECES

Pieces of dust or smoke are needed to help precipitation form. Water vapor condenses around these pieces, which are called condensation nuclei. This allows the tiny drops of water, called droplets, to gather and get heavy enough to fall as rain or snow.

Make the Grade

When the air temperature is at or below freezing, or 32°F (0°C), all the way to the ground, precipitation falls as snow, not rain.

SNOW TIME!

Some clouds are colder than others. For example, a cloud higher in the sky would be colder than one close to the ground. If it's cold enough, the condensed water droplets freeze. This results in tiny bits of ice, called ice crystals, that can fall as snow.

Make the Grade

As they fall, as many as 200 ice crystals can come together to form a single snowflake.

Believe it or not, most rain starts as snow. Precipitation forms high up in the clouds where it is much colder than on the ground. As the snow falls toward Earth through warmer air, the snowflakes turn to rain drops.

Make the Grade

Every snowflake is one-of-a-kind, but most have six points or sides.

WIDE RANGE OF WATER

Depending on where you live, you may see lots of precipitation or none at all. Some places in the **tropics** get more than 900 inches (22.9 m) of rain per year! Deserts get very little rain. Some get less than 0.1 inch (2.5 mm) per year.

Make the Grade

The city of Iquique, Chile, once went 14 years with no rain!

Mawsynram, in the country of India, holds the record for wettest spot in the world. Between 1989 and 2010, the town had an average of 494 inches (12.5 m) of rain per year. That's more than the height of a telephone pole!

Make the Grade

Mount Waialeale, Hawaii, is the wettest spot in the United States. It gets an average of about 373 inches (9.5 m) of rain per year.

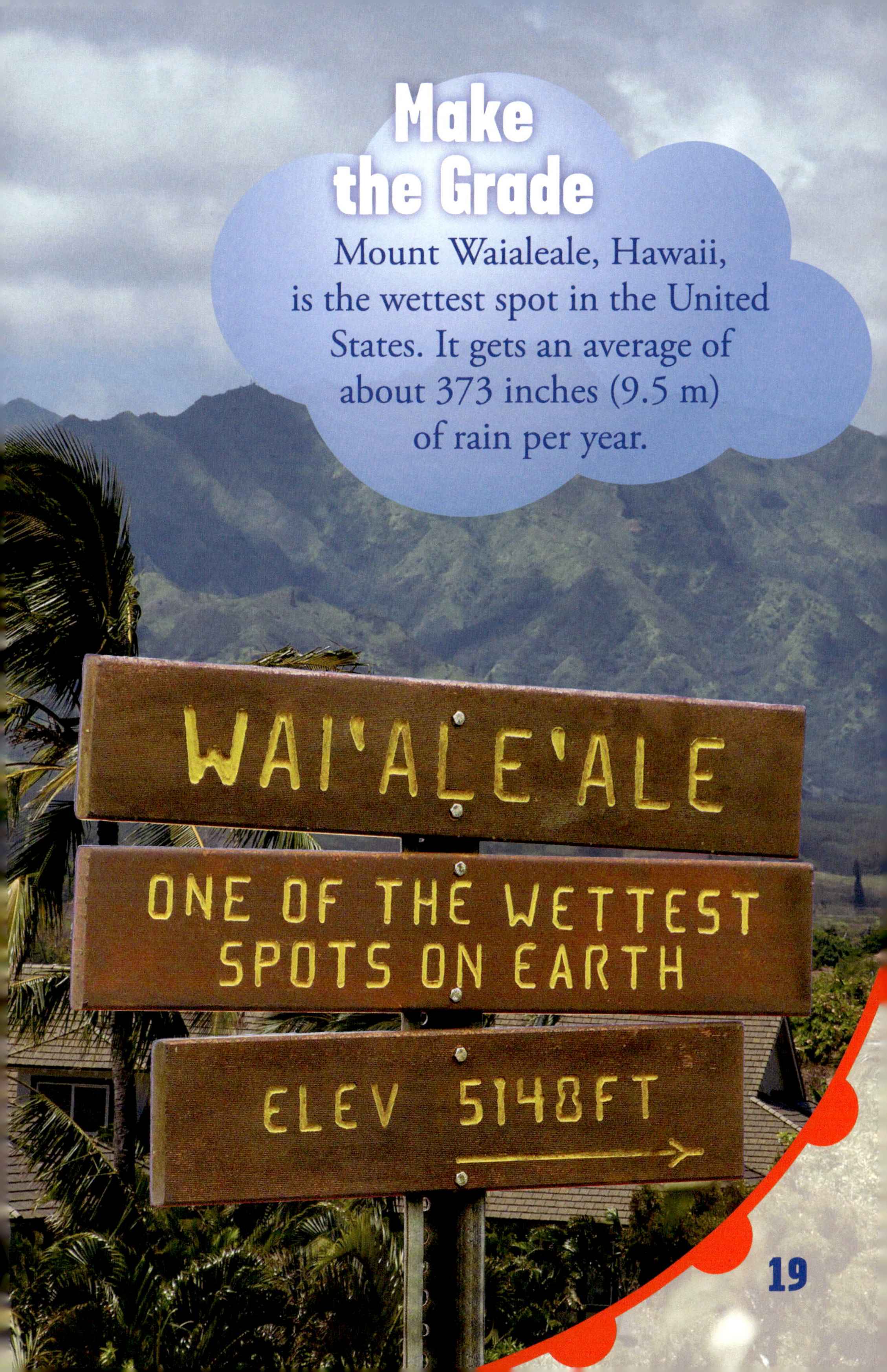

WHERE THE SNOW FALLS

Like rain, snow falls more often in certain places than others. In most places that get snow, it happens mainly during the winter months. But at the North and South Poles, snow covers the ground all year long.

Make the Grade

Near the **equator**, where temperatures are higher than other places on Earth, snow only happens at the top of the highest mountains.

ALL ABOUT AQUIFERS

When precipitation goes through soil and gathers under the ground, it's called groundwater. An aquifer is a layer of rock, sand, or **gravel** that holds groundwater. Much of the world's drinking water comes from aquifers. Farmers often use water from aquifers on their crops too.

Make the Grade

Wells are manmade holes used to reach the water in aquifers.

ACID RAIN

Pollution in the air can get into water droplets before they fall to Earth. This precipitation is known as acid rain. It can pollute the bodies of water the rain falls into. This can hurt the plants and animals that live in or near these waters.

Make the Grade

Pollution in acid rain often comes from smoke from cars or factories.

Acid rain can happen anywhere, even in places with very little or no pollution. This is because wind carries pollution all over the world. Governments from all countries need to work together to control pollution and acid rain.

Make the Grade

Acid rain can even cause parts of buildings to wear away.

THE WORLD'S WATER

The water cycle keeps the world running. Rain and snow are important parts of the water cycle. Farmers depend on rain to keep their crops and animals healthy. Everyone needs rain to have a ready supply of drinking water and food to eat.

Make the Grade

Precipitation often starts with water from the ocean, but rain and snow are never salty. This is because when seawater evaporates, the salt stays behind.

THE WATER CYCLE

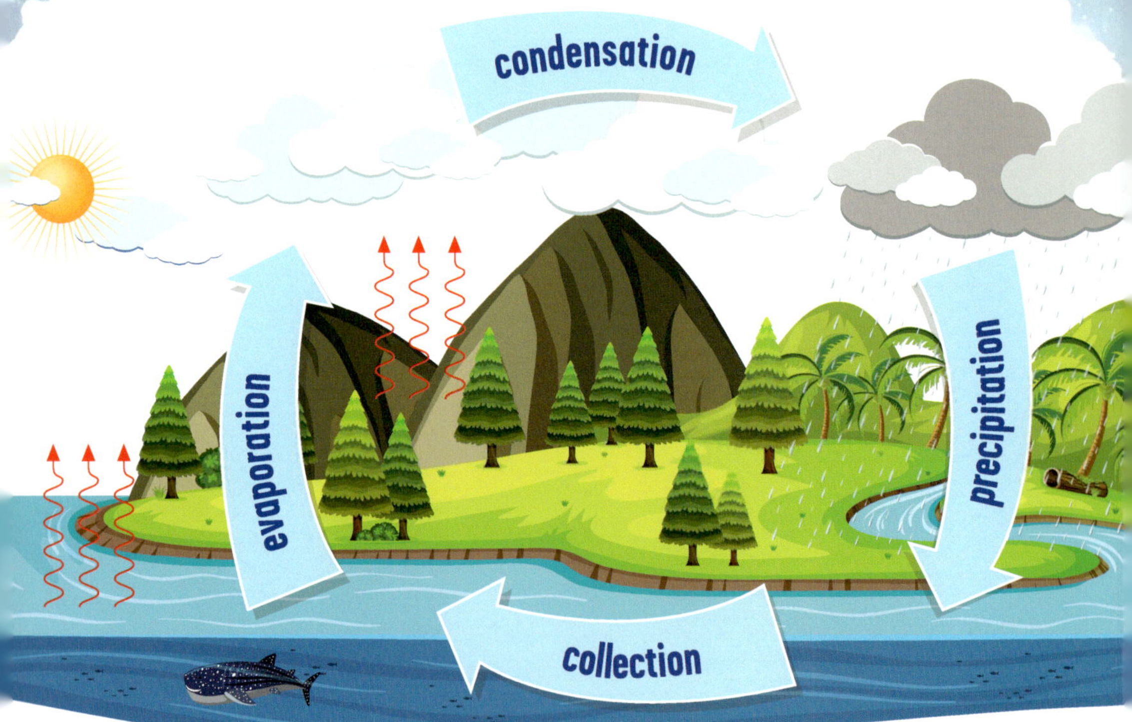

GLOSSARY

atmosphere: The mixture of gases that surround a planet.

equator: An imaginary line around Earth that is the same distance from the North and South Poles.

evaporate: To change from a liquid to a gas.

gravel: Loose pieces of rock.

hail: Small lumps of ice that sometimes fall from clouds.

liquid: Something that flows and takes the shape of the container holding it.

pollution: Something made unclean, especially by human-made waste.

solid: Something that holds its size and shape.

tropics: The area surrounding the equator.

FOR MORE INFORMATION

BOOKS

Farndon, John, Sean Callery, and Miranda Smith. *Weather.* New York, NY: Scholastic, 2020.

Peterson, Megan Cooley. *Understanding Weather*. North Mankato, MN: Pebble, 2021.

WEBSITE

Precipitation
education.nationalgeographic.org/resource/precipitation
Find out more about rain, snow, and other forms of precipitation here.

INDEX